KU-250-550

The Pebble
Spotter's
Guide

Previous page:
A ring of sandstone
pebbles with veins
of quartz.

The Pebble Spotter's Guide

CLIVE MITCHELL

Ilustrated by
Ella Sienna

I dedicate this book to my beautiful wife Joy.
We have shared many happy times on beaches, with the occasional
pebble picked up along the way, and look forward to many more.

Published by National Trust Books
An imprint of HarperCollinsPublishers
1 London Bridge Street
London SE1 9GF
www.harpercollins.co.uk

HarperCollinsPublishers
Macken House
39/40 Mayor Street Upper
Dublin 1, D01 C9W8, Ireland

First published in 2021
Reprinted with changes 2022

Copyright © HarperCollins Publishers Ltd 2021
Text copyright © Clive Mitchell 2021
Illustrations © Ella Sienna

The moral rights of the author have been asserted.

All rights reserved. No part of this publication may be copied, displayed, extracted, reproduced,
utilised, stored in a retrieval system or transmitted in any form or by any means, electronic,
mechanical or otherwise including but not limited to photocopying, recording, or scanning
without the prior written permission of the publishers.

The contents of this publication are believed correct at the time of printing. Nevertheless, the
publisher can accept no responsibility for errors or omissions, changes in the detail given or for
any expense or loss thereby caused.

National Trust edition   ISBN 9781911657514   10 9 8 7 6 5 4
Trade edition   ISBN 9781911657309   10 9 8 7 6 5

A CIP catalogue record for this book is available from the British Library.

Reproduction by Rival Colour Ltd, UK
Printed in Slovenia

If you would like to comment on any aspect of this book, please contact us at the above address or
national.trust@harpercollins.co.uk

National Trust publications are available at National Trust shops or online at nationaltrustbooks.co.uk

Thanks to Eunice Buchanan and Black & White Publishing Ltd for permission
to use the extract from 'Just Another Pebble' on p. 89

MIX
Paper | Supporting
responsible forestry
FSC
www.fsc.org
FSC™ C007454

Front cover illustration National Trust edition 'Pebbles'
by Diana Leadbetter courtesy of Yellow House Art
Licensing www.yellowhouseartlicensing.com
Front cover illustrations Trade edition © Ella Sienna

Contents

Introduction   6

A–Z of Pebbles   12–101

    Features

    The Secret of Finding Pebbles 24

    The Art of Pebbles 40

    Magical Pebbles 56

    Skimming Stones 74

    Pebble Poetry 88

Glossary 102

Index 108

Dimensions of sample pebbles 110

Recommended guides 111

Acknowledgements 112

Introduction

They say that pebble-hunting is what geologists do on holiday, and my first experience of geology was over 50 years ago, picking up pebbles on the beaches of Cornwall and Devon on family holidays. I had no idea what they were. It was purely the tactile pleasure of holding a perfectly smooth pebble that fitted neatly into the palm of my hand. So while a more technical definition of a pebble is any rock between 4 and 64mm (3/16–2^{1}/2in) in diameter, mine is simply:

"A pebble is a smooth rock that fits neatly into the palm of your hand."

This of course means that pebbles will range in size depending on how large your hands are – and that is part of the point – the pleasure of a pebble is personal. The best pebbles are always the ones that you find yourself, that appeal to your own preferences for colour, texture and shape, and slip into your hand as if they were designed just for you.

As a professional geologist I like to know what my pebbles are made of, and I can usually work it out on the spot. For the non-geologist this will be a little harder as rocks just look like, well, rocks.

Pebbles are usually, but not always, formed from a naturally occurring rock that has been worn smooth by the action of water on beaches, or in lakes and rivers. There are also pebbles formed from artificial materials such as concrete, brick and glass; while these are not rocks, they often make interesting pebbles that are sometimes hard to distinguish from rocks. Older forms of concrete can often look like a pebbly sandstone.

A few basics to set the scene, geologically speaking. I have been referring to pebbles as rocks. You might have been calling them stones, but this technically only refers to rocks that have 'stone' in their name such as limestone, sandstone, gritstone, siltstone and mudstone. If you start using the word 'rocks' instead of stones, you'll be in my good books.

Gneiss

Naturally occurring rocks fall into three categories:

- **Igneous:** these are rocks that have formed from molten magma, ranging from those created by volcanic eruptions, such as basalt and andesite, to those with large crystals formed by slow cooling underground, such as granite and gabbro.

- **Sedimentary:** these are rocks formed by the processes of weathering, erosion and deposition, including sandstone, siltstone and mudstone, or made from the remains of past life, such as chalk and other forms of limestone.

- **Metamorphic:** these are rocks formed when other rocks are changed by intense heat or pressure, or a combination of both. So mudstone is squashed into slate and ultimately the minerals are recrystallised to form schist. Limestone becomes marble, with all evidence of past life gradually removed. Granite takes on a whole new appearance when the minerals it contains, such as quartz, feldspar and mica, are reshaped into the swirling layers of gneiss.

Rocks are made of minerals. Identifying these minerals is the first step in working out what a rock is. This is easy when the minerals form the lovely large crystals that you often see in granite. But many rocks are made of very small mineral grains or crystals that are often impossible to see without a magnifying glass or microscope. But don't despair – even professional geologists struggle to identify these without a laboratory. Other aspects of rocks used in their identification include colour, the size of the mineral grains or crystals, and the layers and other textures.

To help you in your quest to identify your pebble there are many good guidebooks on rocks and minerals (see the recommended list at the end of this guide). One of my favourite methods is to find a pictorial guide and compare the images with what I've found – it works for me!

I hope that the copy of *The Pebble Spotter's Guide* you hold in your hand will point you in the right direction and help you to work out what your pebble is made of. If not, you can always send me a photo at cjmi@bgs.ac.uk or on Twitter at @CliveBGS and I'll do my best to work out what it is for you.

Good luck and happy pebble hunting!

Clive Mitchell
CHARTERED GEOLOGIST (CGEOL), BRITISH GEOLOGICAL SURVEY

Collecting pebbles

It's usually fine to take one or two pebbles for your personal collection, but check the local byelaws and follow all local signage. The best advice is always:

"Leave only footprints. Take only photos."

Photographs are great for recording your finds and sharing them online. Many smartphones will also allow you to save GPS data with the image file to record the location.

And, of course, be sure to write the details in this book to save them for the future.

Cracking pebbles

Cracking a pebble open with a hammer and chisel can expose the fresh rock surfaces within and even reveal an ammonite (a mollusc fossil) or a geode (a hollow stone partially filled with crystals). This experiment is more for the amateur geologist than the general pebble enthusiast, and strictly for adults only.

Staying safe

- Stay away from cliff edges, and the base of cliffs too: rock falls can happen at any time.

- Do not climb or walk over landslide or rock-fall debris, especially after wet weather.

- Always pay attention to warning signs; they are there to advise you on how to stay safe.

- Check the weather forecast and tide times before you go. The sea comes in and out twice a day – the times vary – and it is possible to get cut off by the incoming tide or forced up against the cliffs.

- Beware of steep, shelving beaches and large waves.

- If you are looking for fossils, only collect them from the beach. Do not hammer into the cliffs or solid rock as this will cause long-lasting damage and can be dangerous. Always wear safety goggles and gloves if using a hammer and chisel.

Alabaster

This is a lovely pinkish-white, finely crystalline pebble. Alabaster is a sedimentary rock that is mainly composed of the mineral gypsum, which is soft and easily carved. However, it also dissolves in water, so sculptures or other objects made of alabaster can only be found indoors. You'll often come across alabaster monuments inside churches or cathedrals.

Alabaster is white to off-white in colour with shades of yellow, grey, red and brown. As gypsum is very soft (2 out of 10 on the Mohs scale of hardness, see p. 105) it can be scratched with a fingernail, which is how to tell it apart from the slightly harder pale-coloured limestone.

Alabaster is beautifully translucent: if you hold up a piece to the light, some will shine through it. Thin sheets of alabaster were used to create windows such as those in Valencia Cathedral in Spain.

Found: Penarth Beach, Vale of Glamorgan, Wales

I SPOTTED THIS PEBBLE

AT

ON

Amygdaloidal Basalt

This dark pebble is covered with white spots like a galaxy of stars and planets. It is amygdaloidal basalt, an igneous rock formed by volcanic lava that was rich in gases that failed to escape before the rock cooled and solidified. These bubbles of gas formed small cavities in the lava known as vesicles, which tend to be more concentrated in the upper part of the lava flow. These holes, when filled with minerals such as calcite, chert, quartz and zeolite, are known as amygdales.

Basalt is dark grey to black in colour, and weathers to a reddish or greenish crust with white patches. One of the most famous basalt sites in the world is Northern Ireland's amazing Giant's Causeway, which is made up of 40,000 interlocking basalt columns.

Found: Cayton Bay Beach, North Yorkshire, England

I SPOTTED THIS PEBBLE

AT ....

ON ....

Basalt

This is a hard, dense dark grey pebble of basalt with some very small holes (known as vesicles). The commonest volcanic rock on the planet, basalt is an igneous rock formed from cooling lava. Once the magma erupts on to the earth's surface, it cools quickly, so the crystals remain small as the rock solidifies before they can grow larger.

In some places the lava forms columns with a hexagonal cross-section, a phenomenon known as columnar jointing. As well as at the Giant's Causeway in Northern Ireland, this can be seen at the Devils Tower in Wyoming, USA, and the Canary Islands off the coast of Africa.

Basalt is dark grey to black in colour and weathers to a reddish or greenish crust. It is a hard rock that feels heavier than other rocks of a similar size. It is fine-grained with crystals too small to be seen without the use of a magnifying glass.

Found: Giant's Causeway, County Antrim, Northern Ireland

I SPOTTED THIS PEBBLE

AT ....

ON ....

Brick

Most of the pebbles in this book have been around for ages – literally. But you can pick up some interesting ones that might be younger than you are. Bricks find their way on to beaches from buildings that have fallen into the sea due to erosion undermining coastal cliffs. A typical two-storey house uses some 12,000 bricks.

Brick pebbles are reddish to dark brown, orange, yellow, buff or black in colour – this one is red with yellowish-brown mottling. You might mistake brick for sandstone or a volcanic rock, but bricks are generally more porous, friable (a geologist's word for crumbly) and lightweight than rock pebbles. Another clue might be some mortar, paint or other evidence of it being used in construction. They could also be marked with the name of the brick factory, which is a bit of a giveaway!

Found: Southwold Beach, Suffolk, England

I SPOTTED THIS PEBBLE

AT ....

ON ....

Chalk

Look at the marvellous uniformity of this off-white chalk pebble, worn smooth by the sea. It's the perfect shape and size for skimming – but what a waste that would be.

Chalk is a sedimentary rock made of the mineral calcite with layers of flint nodules and clay. It's a form of limestone consisting of microscopic fossils known as coccoliths. The chalk you're probably most familiar with is that used in schools for chalkboards, which is actually made from the slightly softer mineral gypsum (Mohs 2). If you used 'proper' calcite chalk (Mohs 3) on chalkboards it would produce a teeth-jarring screech.

Chalk is usually white in colour, but occasionally can be grey, yellow, buff or red. It's a soft material that breaks easily and will leave a powdery white dust on your hands.

Found: Mappleton Beach, East Riding of Yorkshire, England

I SPOTTED THIS PEBBLE

AT ....

ON ....

Chert

This dark-red chert pebble has an interesting fracture running through it, probably caused by smashing against other pebbles on the beach.

Chert is a hard sedimentary rock, very similar to flint. It is made of silica in the form of 'cryptocrystalline' quartz (crystals too small to be seen with the human eye). Chert is a 'biochemical' rock made from the hard parts of plankton and sponges that were deposited on to the deep ocean floor. This sediment solidifies to form beds of chert. It can also form as rounded nodules in limestone.

Chert can be red, black, white, brown, green or grey in colour. It has an almost glassy or ceramic appearance. When broken it has a partially concave surface and, like flint, will splinter to produce very sharp fragments (which were used to make cutting tools in the past).

Found: Hornsea Beach, East Riding of Yorkshire, England

I SPOTTED THIS PEBBLE

AT ....

ON ....

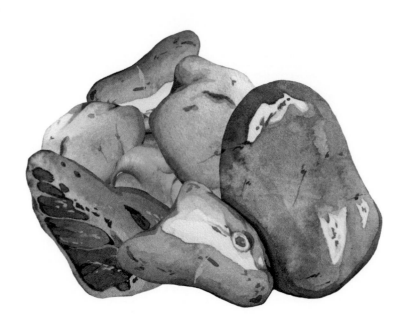

The secret of
finding pebbles

———————

Head down, walking across the shingle, beady eyes rapidly scanning for interesting-looking pebbles … Sound familiar? Not only is this a recipe for a stiff neck, it's also likely that you're missing out on some hidden gems.

A more scientific method I've used before is to place a one-metre (one yard) square grid on a beach and identify every single pebble inside it. I discovered that the pebbles are often sorted vertically, with small, interesting pebbles hidden under the larger ones; the same way that raisins often sit at the bottom of a cereal packet.

So my advice is to slow down, focus and get into the 'pebble zone'; sit yourself down on the beach and go through every pebble within reach. Look at each one properly. Not only will you reconnect with the unique sounds, sights and smells of the shoreline, but you might just uncover that prize pebble.

Coal

We still have trees, so why isn't coal still being created? Well, coal was mainly formed in the Carboniferous period from the buried remains of forests that immense pressure, heat and time transformed into the fuel that still provides much of the world's energy today, despite our efforts to cut back on its use. After the Carboniferous period, however, about 300 million years ago, bacteria evolved that eat wood lignin, so trees didn't survive long enough to be turned into more coal.

Coal is black or dark brown in colour. Our pebble is black coal with a smooth surface and visible layers. It is a relatively soft rock that will break apart and leave black dust on your hands. Its relatively light weight means it is more affected by the ebb and flow of the sea, and will be found along the high-tide mark on a beach.

Found: Staithes Beach, North Yorkshire, England

I SPOTTED THIS PEBBLE

AT ....

ON ....

Concrete

Don't mistake those flecks for ancient fossils or minerals. This is a whitish-grey concrete pebble containing small, rounded gravel pebbles.

Concrete is, of course, artificial, and is a mixture of sand and fine aggregate bound together with cement. The Romans were the first to use concrete around 2,000 years ago. Rome's Colosseum and Pantheon were built with concrete, and the Pantheon is still the largest unreinforced concrete dome in the world. The Romans discovered that pozzolana, a volcanic sand found near Naples, is a naturally occurring cement that makes a very durable concrete.

Like brick, concrete finds its way on to beaches from buildings that have fallen into the sea due to coastal erosion. It's mostly grey, but can be virtually any colour. It may contain fragments of mortar, paint or other construction material, and can often be confused with sandstone or conglomerate.

Found: Clevedon Beach, North Somerset, England

I SPOTTED THIS PEBBLE

AT ....

ON ....

Conglomerate

In 2012 NASA's *Curiosity* rover found an outcrop of conglomerate on Mars. But you don't have to venture that far to find examples.

Conglomerate is a sedimentary material composed of fragments of any type of rock (known as clasts) set in a finer matrix of sand, silt or clay. It is formed in rivers, lakes and beaches where the action of water has smoothed the rock fragments round.

The clasts can range from 2mm (5/64in) across to boulder-sized. The colour of the pebble will vary depending on the type of rocks that the clasts are made of and the material that binds them together.

There's a type of conglomerate called 'puddingstone', where the clasts look like raisins in a pudding, but our pebble contains rounded clasts that range from 3–4mm to 3–4cm (^{1}/8–1^1/2in) in size, and are light to dark brown, brick-red and purple in colour.

Found: Cushendun Beach, County Antrim, Northern Ireland

I SPOTTED THIS PEBBLE

AT ....

ON ....

Coral Limestone

What a work of art this little rock is. You could look at every one of these light-grey flecks and not find two the same. That's because they are the fossils of an ancient coral colony held within the surrounding darker limestone.

Coral limestone is a sedimentary rock formed from the remains of coral reefs, like the modern-day barrier reefs in the Bahamas and off the east coast of Australia. The coral in this pebble grew in a warm, clear, shallow, well-lit tropical sea over 300 million years ago – not much like today's Yorkshire coast.

Limestone is white, grey, pink, red, cream or black in colour. It's a 'biogenic' rock (composed of the remains of plants and animals) and often contains easily visible fossils, such as the coral in this pebble. Made of the mineral calcite, it's a relatively soft rock (Mohs 3) and easily scratched with a steel knife.

Found: Mappleton Beach, East Riding of Yorkshire, England

I SPOTTED THIS PEBBLE

AT ....

ON ....

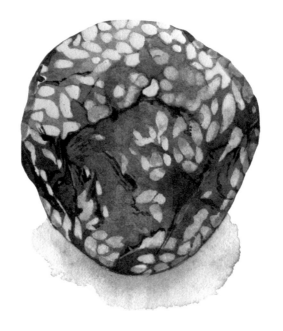

Dolerite

At first sight this might look like a nondescript pebble, but it's steeped in history – this hard bedrock forms the foundation for Hadrian's Wall, and a type of dolerite called bluestone was integral to the construction of Stonehenge.

This is a dark greenish-grey, hard, compact pebble with medium-sized crystals.

Dolerite (otherwise known as diabase) is an igneous rock formed from magma that has pushed its way through cracks and fissures in other rocks. This is called an intrusion, and can either be found cutting across the rock layers (known as a dyke) or parallel to them (known as a sill). Dolerite is made of the minerals feldspar and pyroxene, with smaller amounts of magnetite, olivine and quartz.

Found: Fair Head Sill, Ballycastle, County Antrim, Northern Ireland

I SPOTTED THIS PEBBLE

AT ....

ON ....

Dolostone

One way of identifying a dolostone pebble is that it's hard and feels heavier than limestone pebbles of a similar size. Dolostone is a sedimentary rock formed when magnesium changes the mineral calcite in limestone to dolomite. You'll rarely find fossils in dolostone as this process tends to destroy them.

Dolostone is white, creamy yellow or grey when freshly broken, and weathers to a pale brown or pinkish colour. It has a characteristic sandy or sugary-looking surface texture with small holes (known as pores). Dolostone often occurs wherever you find limestone, a classic location being the Dolomite Alps in northern Italy. The rock was in fact originally known as dolomite, but the name was changed to dolostone in 1948 to reduce confusion with the mineral dolomite – though some geologists objected to this because the rock was called dolomite first.

Found: Brean Beach, North Somerset, England

I SPOTTED THIS PEBBLE

AT ....

ON ....

Echinoid in Flint

Sometimes you find a pebble so marvellous you just keep taking it out and looking at it all day. And this particular one is more common than you might think, so keep your eyes peeled.

This pebble contains the fossil of an echinoid (pronounced 'eck-in-oid'), otherwise known as a sea urchin, preserved in a flint nodule. Echinoid fossils have characteristic pairs of parallel marks that radiate out from the top; these are where the spines of the sea urchin were attached.

The original delicate shell of the echinoid has been replaced by silica in the form of 'cryptocrystalline' quartz, preserving its wonderful starlike appearance. If you find an echinoid fossil in a pebble, you can be reasonably sure that the rock in question was formed in a marine environment.

Found: Cromer Beach, Norfolk, England

I SPOTTED THIS PEBBLE

AT ....

ON ....

The art
of pebbles

Of course, I think pebbles are perfect as they are, but throughout history people have always loved decorating them. Prehistoric painted pebbles dating from over 10,000 years ago have been found in southern France and Spain. The pebbles shown here are small, rounded quartzite beach pebbles that were painted, probably with peat tar, by Picts in the Caithness, Orkney and Shetland regions of Scotland between AD 200 and 800. Whether they were used as slingstones, or charms, or both, is unknown. What seems certain is that they would have been highly prized by their owners.

Flint

Watch yourself around flint. It splinters into very sharp fragments. That's why it was such an invaluable tool for Stone Age people. To this day, the sparks created by flint against steel are used as a fire starter.

Flint is a hard sedimentary rock that commonly occurs as nodules in Cretaceous chalk. Like chert, it is made of silica in the form of cryptocrystalline quartz. It also forms irregular barrel-shaped masses with the marvellous name of 'paramoudra' (nicknamed 'pot stones'), which can be up to 2 to 3 metres (6 to 10ft) across. These really are the weirdest rocks you are likely to see on a beach.

Flint is blue-grey, grey to nearly black in colour, and weathers to a whitish, powdery surface crust. When broken it has an almost glassy appearance and a partially concave surface.

Found: Lyme Regis Beach, Dorset, England

I SPOTTED THIS PEBBLE

AT ....

ON ....

Fossiliferous Limestone

I make no apologies for including another example of limestone. It's useful stuff. Did you know you use a little bit every morning in your toothpaste? Its mild abrasive property helps remove plaque, and the calcium content is good for your teeth.

This is a grey limestone pebble with fragments of crinoid stems and other fossils. Crinoids are marine animals related to the starfish, still living today but massively more abundant in the past. The longest crinoid stem fossil ever found was 40 metres (130ft) long – a bit bigger than ours!

Limestone is a sedimentary rock composed of the mineral calcite. It may also contain dolomite, pyrite, galena, fluorite, quartz or iron oxides. It's a very versatile material with many industrial uses such as building stone, aggregate for roads, and the manufacture of cement. We'd be lost without it.

Found: Clevedon Beach, North Somerset, England

I SPOTTED THIS PEBBLE

AT ....

ON ....

Gabbro

Not to be confused with its exotic cousin indigo gabbro (aka 'Mystic Merlinite', prized by people seeking to hone their psychic powers), this attractive pebble is formed of mottled black and white gabbro with large crystals of white feldspar and dark greenish-grey augite.

Named after a hamlet in Tuscany, Italy, gabbro is a hard igneous rock with large crystals created when magma cools slowly underground. It makes up a large part of the planet's oceanic crust, and can also occur as large intrusions of magma deep underground. These 'lopoliths', as they are known, can be revealed at the surface of the earth by erosion over millions of years.

Gabbro often contains the green mineral olivine. Its pebbles are most often black and white like this one, but can also be grey, dark greenish or bluish.

Found: Whiting Bay, Isle of Arran, North Ayrshire, Scotland

I SPOTTED THIS PEBBLE

AT ....

ON ....

Glass

This is a brownish-grey, smooth and rounded glass pebble. Although human civilisation has been making glass for over 5,000 years, it wasn't until 1959 that modern-day float-glass manufacturing started. Float glass is made by melting quartz sand with feldspar and other minerals. Glass finds its way on to beaches from discarded bottles, glass manufacturing waste or from buildings that have fallen into the sea. The combined action of wind, wave and rubbing against other pebbles soon makes it as smooth as... well, glass.

Glass is mostly clear, green or brown, and occasionally blue in colour. Glass pebbles often have a frosted surface from being tumbled with other pebbles on the beach. They may be transparent and have a partially concave surface when broken.

Found: Seaham Beach, County Durham, England

I SPOTTED THIS PEBBLE

AT ....

ON ....

Glauconitic Sandstone

This pebble's glory is its deep, olive-green colour.
Try to ignore the fact that the colour comes from
fossilised dung...

Glauconitic sandstone (also known as 'greensand')
is a sedimentary rock mostly made of pellets of the
green clay mineral glauconite with some quartz
and clay. Glauconite is formed from the fossilised
faeces of trilobites and other animals that lived at the
bottom of the sea. It's been used as a green pigment
in paintings since Roman times, and also as a
fertiliser because of its high potassium content.

Glauconitic sandstone is typically olive green but
may be red, brown or grey in colour. Its surface can
be rough to the touch or smooth like clay, and it may
have layers or fine laminations.

Found: Hornsea Beach, East Riding of Yorkshire, England

I SPOTTED THIS PEBBLE

AT ....

ON ....

Gneiss

A pebble like this might just be the oldest thing you've ever held in your hand.

Gneiss was mostly formed in the Precambrian geological time period and often makes up the 'basement' that all other rocks sit on. The oldest rock in the world is thought to be the gneiss in Quebec, Canada – over 4 billion years old.

Gneiss is pronounced 'nice', which can lead to some confusing conversations. Turn your pebble over in the light, and you'll see how it glistens – its name comes from an old word meaning 'spark'.

It's a hard metamorphic rock that started life as sedimentary or igneous rocks. Transformed by high temperatures and pressures deep underground, these recrystallised into gneiss. Our pebble has alternating layers of minerals; the pale layers contain feldspar and quartz, and the darker layers mica and hornblende. There's a lovely pink example on p. 7.

Found: Newtonmore, Highland, Scotland

I SPOTTED THIS PEBBLE

AT ....

ON ....

Gryphaea

Imagine picking this beauty up off a beach without knowing a thing about geology – you'd wonder what strange beast it had fallen off. It's no wonder this became known as a Devil's Toenail!

It's actually a fossil of a gryphaea preserved in grey limestone. A 'biogenic' rock (see p. 102), limestone can be white, grey, pink, red, cream or black in colour, and often contains easily visible fossils. It is a relatively soft rock as it is made of the mineral calcite, and is easily scratched with a steel knife.

Gryphaea are an extinct form of oyster, but the characteristic curved shape (some are much more curved than the one I found – have a look online) and strong ridges certainly have the look of the grisly by-product of the Devil's attempts at a pedicure. They were also thought to be a cure for arthritis.

Found: Cayton Bay, North Yorkshire, England

I SPOTTED THIS PEBBLE

AT ....

ON ....

Magical pebbles

Hag stones (also called adder stones, witches' stones and serpents' eggs) are rocks – usually but not always flint – with naturally occurring holes. Whatever you call them, they're weirdly beautiful and are supposed to be lucky, so don't throw them away. There are several theories as to how hag stones form: they come from the sting of an adder; they are the hardened saliva of a group of serpents; they result from small shellfish boring a hole; they're formed by the action of waves finding a weakness in the rock.

And if you think the first two ideas sound far-fetched, another mind-blowing but scientifically sound theory is that sometimes the hole was there first and the rock came later. Animals now long-extinct supposedly burrowed in chalk to form a chamber that was then slowly partially filled by flint nodules. What wasn't filled remained a hole.

Hornfels

This has to be one of the strangest rocks around. Known as a 'ringing rock' because it sounds like a bell when struck with a hammer, hornfels is a metamorphic rock formed by the intense heat of an igneous intrusion changing rocks such as shales and clays, limestones and igneous rocks. Its musical quality means it can be used to make a 'lithophone' (a stone xylophone). The Lake District's Till family gave concerts on such instruments in Europe and America in the 19th century; they were the original 'rock band'!

Hornfels is black or bluish- or greenish-grey in colour and has very fine-grained crystals. It is often speckled with porphyroblasts of minerals such as andalusite, cordierite, garnet and sillimanite. Faint sedimentary layers from the original rock may remain. It is notoriously hard to break.

Found: Mourne Mountains, County Down, Northern Ireland

I SPOTTED THIS PEBBLE

AT ....

ON ....

Ironstone

Despite the name, this rock plays no part in the production of ironstone china, which is so called only for its hardwearing quality.

Ironstone is made of a sedimentary rock such as sandstone or limestone that contains more than 15 per cent iron in minerals such as hematite. It was a valuable source of iron during the Industrial Revolution, but is no longer used for this as other, more economical sources have been discovered.

Ironstone is dark red, rusty brown or yellow in colour. It will often be medium-grained i.e. have sand-sized particles. Our particularly tactile pebble is a distinctive rusty orange-brown with a sandy texture and bluish-black layers.

Ironstone feels heavier than other pebbles of sandstone or limestone of a similar size, and often occurs as hard, rounded nodules. It may also contain fossils of shellfish or plants.

Found: Whitby Beach, North Yorkshire, England

I SPOTTED THIS PEBBLE

AT ....

ON ....

Jet

I had spent years searching for jet until I found this unassuming little pebble, which I nearly ignored. A lesson to examine everything.

Whitby is famous for its jet, which is thought to be fossil wood from a tree that is related to the modern-day Monkey Puzzle tree. Jet was made popular by Queen Victoria, who wore polished jet jewellery in mourning for Prince Albert after his death in 1861, although it has been used in jewellery in Britain from Roman times.

Jet is black to dark brown in colour and warmer to the touch and lighter in weight than other rocks. Freshly broken pieces have a distinctive oily smell. It can be found washed up on a beach on an ebbing tide, but can easily be confused with coal. To tell them apart, drag them across a hard surface such as concrete; jet will leave a brown line and coal a black one.

Found: Whitby Beach, North Yorkshire, England

I SPOTTED THIS PEBBLE

AT ....

ON ....

Lapilli Tuff

This lapilli tuff is made of volcanic ash ('tuff')
with dark grey to black teardrop-shaped fragments
('lapilli') of volcanic glass and rock.

Lapilli tuff is a crossover between an igneous and a
sedimentary rock. The volcanic ash is deposited like a
sandstone and may be termed a 'volcaniclastic' rock.
The lapilli (Latin for 'little stones') were molten
when they were erupted from the volcano. As they
travelled through the air they became streamlined
and cooled to form wispy strands of glass. Generally,
the larger a piece is, the closer it fell to the volcano
that expelled it; most fall within 20km (12 miles) of
the eruption.

Lapilli tuff varies in colour: light to dark brown,
grey to black, occasionally pinkish, yellowish or
greenish. It has distinctive wispy fragments of darker
volcanic glass.

Found: In a stream, Patterdale, Lake District, England

I SPOTTED THIS PEBBLE

AT ....

ON ....

Layered Sandstone

Another rock that repays closer inspection. Look at those almost imperceptible laminations, caused by tiny differences in the varying sediment layers that have contributed to this rock's geological journey.

This is a pebble of pale rusty-brown rounded sandstone with a sandy surface and those fine laminations.

Sandstone is a sedimentary rock, typically made of quartz sand (though it can be any mineral), that can occur in many different types of geological environment such as the sea, lakes, rivers and deserts. Beach sand can also turn into a sandstone known as 'beach rock'. This can be comparatively young for a rock, at less than 2,000 years old.

Found: Hornsea Beach, East Riding of Yorkshire, England

I SPOTTED THIS PEBBLE

AT ....

ON ....

Limestone

At first glance, it looks like someone's left a boot print on this pebble, but the black pattern is actually the fossil of a shellfish. Limestone is sedimentary, and one of the most commonly occurring rocks on the planet. As I mentioned earlier, it is a 'biogenic' rock, composed of the remaining hard parts of plants and animals.

The fossil in this pebble is a small marine shellfish known as Plagiostoma, and could be over 200 million years old. It has a hinged shell in two parts – what is referred to as a bivalve. Plagiostoma would have lived in a sea that was teeming with ammonites and ichthyosaurs – a veritable prehistoric soup.

Found: Penarth Beach, Vale of Glamorgan, Wales

I SPOTTED THIS PEBBLE

AT ....

ON ....

Phyllite

You have to keep your eyes peeled when pebble spotting. It would be all too easy to dismiss this as just another grey pebble, but look closely at its beautiful wavy surface and feel that texture. Gorgeous.

As well as the waves, this steel-grey, flat pebble of phyllite (pronounced 'phil-ite', from the Greek *phyllon*, meaning 'leaf') has small crystals known as porphyroblasts dotted across its surface.

Phyllite is formed when mudstone is transformed by heat and pressure, and consists mostly of the minerals mica and quartz with chlorite and graphite, and porphyroblasts of andalusite or cordierite. The mica is too small to see without a magnifying glass (if the mica is large enough to see then what you're holding is probably a schist).

Like slate, phyllite splits easily into sheets of rock used as roofing tiles, although they tend to be thicker than slate tiles.

Found: Killegruer Beach, Kintyre, Argyll & Bute, Scotland

I SPOTTED THIS PEBBLE

AT ....

ON ....

Pink Granite

This is one of those pebbles that a splash of water just brings to life, as you'll see from the illustration. It looks like it has just been plucked from the sea.

It's a granite pebble with salmon-pink feldspar, glassy-grey quartz and black biotite mica from the breathtaking Mourne Mountains in Northern Ireland. The granite there was formed around the same time as the basalt at Giant's Causeway, 150km (95 miles) to the north.

Granite from the Mourne Mountains has been used to pave the streets of many cities, including Liverpool and Manchester, as a building stone, for example in the Northern Ireland parliament building at Stormont, and in monuments around the world, such as the Queen Elizabeth II September 11th memorial garden in New York.

Found: Murlough Beach, Dundrum, County Down, Northern Ireland

I SPOTTED THIS PEBBLE

AT ....

ON ....

Skimming stones

As a kid I was obsessed with skimming stones on the beach. As an adult I still can't resist, the only difference being I don't throw any prize specimens in by mistake nowadays.

It starts with wandering along a seashore or lakeside to find a perfect flat, round pebble, then hooking it up to your index finger, getting nice and low to the ground and, with a rapid flick of the finger, sending it skipping and skimming over the surface of the water. The best rocks for skimming are flat, obviously, 5–10mm ($^1/_4$–$^1/_2$in) thick and 5–10cm (2–4in) across; apparently 20 degrees from the horizontal is the optimal angle to aim for when you launch one. Some rocks naturally form flat pebbles: sandstone, shale, siltstone, slate, limestone and schist in particular.

See how many times you can make a pebble bounce – an incredible 88 times is the record.

Quartz Porphyry

It's not that hard to tell the difference between quartz porphyry and rhomb porphyry – the latter has much more distinctive regular-shaped crystals. But there's still something quite entrancing about the clear quartz phenocrysts in this volcanic rock.

Porphyry was much prized by the ancient Romans for its purplish hue, which coincided with the Imperial colour – the name porphyry comes from the Greek word for purple.

Our pebble is a quartz porphyry with phenocrysts of feldspar and quartz set in a light-brownish grey igneous volcanic rock.

Quartz porphyry is generally a light-coloured (red, brown or green) volcanic rock that consists of quartz, feldspar and mica with phenocrysts of quartz and feldspar. The distinctive large crystals of clear quartz are the tell-tale sign of this rock, with larger phenocrysts of white feldspar.

Found: Whiting Bay Beach, Isle of Arran, North Ayrshire, Scotland

I SPOTTED THIS PEBBLE

AT ....

ON ....

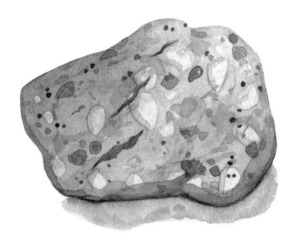

Quartzite

Quartzite is a bit of a Marmite pebble – you'll either be underwhelmed by its bland appearance or dazzled by its creamy clarity. But don't write it off too quickly – you can find examples with lovely subtle pink, red, yellow and orange discolouration. It also forms satisfyingly smooth and rounded pebbles that will take a beautiful polish.

This quartzite pebble consists almost entirely of the mineral quartz with a small vein of black mica. It's a hard rock (quartz is 7 on the Mohs scale) formed by the metamorphism of a sandstone, commonly found in areas where the rocks are very old (Precambrian).

Its toughness made it useful for tools in prehistoric times, and it's very resistant to weathering, so is often found on ridges and hilltops. The lack of vegetation is a clue to its presence below a thin layer of soil.

The town of Quartzsite in the desert of western Arizona gets its name from the quartzites found in the nearby mountains.

Found: Killegruer Beach, Kintyre, Argyll & Bute, Scotland

I SPOTTED THIS PEBBLE

AT ....

ON ....

Red Granite

Granite is one of the hardest materials in the world. And yet even this incredibly tough rock is no match for the sea. Its constant ebb and flow causes granite rocks to tumble together, wearing away the rough edges to produce smooth, rounded pebbles.

Granite is an igneous rock resulting from magma cooling slowly underground to form large crystals of quartz, feldspar and mica. It can be a variety of colours including off-white, grey, pink or red, with black spots. It's common, but one of my favourites.

This pebble is formed of large crystals of glassy grey quartz, together with both white and reddish-orange feldspar, and black mica. The crystals in granite can range in size from 2 to 16mm ($5/64$–$1/2$in), with some extra-large ones (known as 'phenocrysts') up to 5cm (2in) long.

Granite is found all over the world. There's a Pink Granite Coast in northern France, and even a village in Wisconsin, USA, called Redgranite!

Found: Craignure Beach, Isle of Mull, Argyll & Bute, Scotland

I SPOTTED THIS PEBBLE

AT ....

ON ....

Rhomb Porphyry

This is my very favourite pebble – it really is a diamond (or should that be rhomb!) – and is the holy grail for some pebble hunters. It's a rounded pebble of rhomb porphyry with pale reddish-brown phenocrysts of feldspar set in a dark-brown volcanic rock.

Rhomb porphyry is an igneous rock that is only known in three places in the world – Mount Kilimanjaro in Tanzania, Mount Erebus in Antarctica and Oslo in Norway. This pebble is thought to have been carried over to the UK from Norway by glaciers during the Ice Ages.

Rhomb porphyry pebbles can be grey, green, red, pink, purple or brown in colour. They have characteristic large diamond-shaped crystals of feldspar. Pale in colour, these often have sharp edges and occupy about half of the rock. Other similar-looking pebbles of porphyry do not have rhomb-shaped phenocrysts.

Found: Cromer Beach, Norfolk, England

I SPOTTED THIS PEBBLE

AT ....

ON ....

Sandstone

If you can tear yourself away from watching for the elegant dolphins that often make an appearance off the shore near Inverness in Scotland, you might find another of my favourite rocks. This is a pebble of reddish-brown sandstone mostly made of medium to coarse-sized sand grains of the mineral quartz.

Sandstone is a sedimentary rock that can be made of any mineral but is typically composed of quartz grains cemented together. It may also contain feldspar and rock fragments, and other minerals such as mica, clay, organic matter, zircon, tourmaline and rutile. The important factor is the size of the sand particles, which can be up to 2mm (5/64in) in diameter.

Sandstone occurs in all sorts of colours including white, grey (see p. 1), green, yellow, cream, buff, brown or red. It often has a rough sandy surface but may also form smooth, hard pebbles of fine sand. It may have layers or fine laminations.

Found: Chanonry Point Beach, Inverness, Scotland

I SPOTTED THIS PEBBLE

AT ....

ON ....

Schist

If you got hold of a piece of mud and squeezed it hard – *really* hard – over millions of years, you might end up with some schist. I wouldn't recommend it, though. Much easier to go out and find some schist pebbles lying around.

All that squeezing recrystallises the clay into the mineral mica. Mica is known as a 'platy' mineral, as it easily breaks into thin transparent sheets. In the past, this see-through mineral was used to make windows, including in old cooking stoves.

Schist is a metamorphic rock, mostly grey in colour with silver, green or blue shades and a very shiny surface thanks to the mica. Our pebble is silvery grey with the minerals quartz, feldspar and mica. It's useful as a building material, for walls or paving.

Found: River Spey, Newtonmore, Highland, Scotland

I SPOTTED THIS PEBBLE

AT

ON

Listen! You hear the grating roar
Of pebbles which the waves draw back, and fling
At their return, up the high strand,
Begin, and cease, and then again begin,
With tremulous cadence slow, and bring
The eternal note of sadness in.

FROM 'DOVER BEACH' BY MATTHEW ARNOLD

How happy is the little stone.
That rambles in the road alone,
And doesn't care about careers

FROM 'SIMPLICITY' BY EMILY DICKINSON

Though the great song return no more
There's keen delight in what we have:
The rattle of pebbles on the shore
Under the receding wave.

WB YEATS, 'THE NINETEENTH CENTURY AND AFTER'

Pebble poetry

---

sometimes

i feel
i could just break out
in rhomboidal
planes & polyhedra

if i weren't
rubbing shoulders
with this lot.

FROM 'JUST ANOTHER PEBBLE' BY EUNICE BUCHANAN

Septarian Concretion

This is a geological lucky dip. Carefully crack this pebble open to reveal beautiful radiating crystals, a perfectly preserved ammonite... or just solid rock. Our pebble is formed of tan brown sandstone with radiating cracks infilled with white mineral.

Septarian concretions (aka septarian nodules) are sedimentary and formed from the same material as the host rock. Minerals like calcite, quartz or iron oxide cement the rock together, which makes it harder and able to survive intact after the host rock has been eroded away.

Septarian concretions are black, brown, grey or yellow in colour. They often form spherical or irregular rounded shapes and may have zigzag-shaped radiating cracks on the surface infilled with a crystalline mineral.

Found: Cayton Bay Beach, North Yorkshire, England

I SPOTTED THIS PEBBLE

AT ....

ON ....

Serpentinite

With its smooth, slippery surface and mottled snakeskin-like pattern, it's easy to see how this pebble gets its name. I wish you could reach into the book and touch it, but you'll just have to find your own. Rather appropriately, this was found at the Lizard in Cornwall.

Serpentinite is a metamorphic rock that is mostly found deep under the ocean, as it forms part of the oceanic crust. In some places, this crust has been forced up on to the land, exposing the layers in a sequence known as an ophiolite. This can be seen in the Semail Ophiolite in Oman and the United Arab Emirates, the Coast Range Ophiolite in California, and the Lizard.

Serpentinite is often used as a decorative stone because it's easy to carve and polish. It can be grey-green, dark green to black in colour, with bright red and yellow patches.

Found: The Lizard Beach, Cornwall, England

I SPOTTED THIS PEBBLE

AT ....

ON ....

Stigmaria

If you happen to find one of these pebbles, lightly run your fingertip over the little reddish-brown circular depressions that pit the surface. These are the rootlet scars of a mighty tree that grew over 300 million years ago.

Lepidodendron were trees that grew to over 30 metres (100ft) in height during the Carboniferous period. They were one of the main plants in the swamp forests that led to the formation of the coal deposits that fuelled the Industrial Revolution of the 19th century.

Stigmaria are pebbles that have preserved their fossil roots in sandstone. The sandstone may be white, grey, green, yellow, cream, buff, brown or red in colour, and often has a sandy surface that varies in roughness.

Found: Runswick Bay, North Yorkshire, England

I SPOTTED THIS PEBBLE

AT ....

ON ....

Tarmac

A deep sniff might be a good way of identifying this pebble, because tarmac pebbles sometimes retain a distinctive bituminous smell. If it's large enough there might be traces of road-marking on it as well.

This is a black, rounded pebble of tarmac (otherwise known as asphalt) with light-grey aggregate of fine-grained rock.

Used to pave roads, tarmac is artificial and made of aggregate (usually crushed rock) bound with bitumen (also known as tar), a sticky black product of oil refining. Typically black or dark blackish-grey in colour, it contains small pieces of paler-coloured aggregate. The bitumen may be slightly pliable.

Tarmac finds its way on to beaches from roads that have fallen into the sea as a result of coastal erosion.

Found: Hornsea Beach, East Riding of Yorkshire, England

I SPOTTED THIS PEBBLE

AT ....

ON ....

Tuff

Look at the intriguing vein running across the top of this pebble, like a river flowing through time itself. This is tuff (pronounced 'tough'), with fine feldspar phenocrysts and an irregular vein of light and dark grey-coloured mineral. Mostly light to dark brown or grey to black in colour, tuff may occasionally have a pinkish, yellowish or greenish hue.

Tuff is a mixture of tiny fragments of glass, mineral and rock erupted from a volcano. It's a crossover between an igneous and a sedimentary rock, as volcanic ash is deposited like a sandstone, which it can resemble. As such, it may be called a 'volcaniclastic' rock.

Such small particles can have a big impact, though. The ash from the 2010 eruption of the volcano Eyjafjallajökull in Iceland grounded many airlines and affected the travel of over 10 million people.

Found: Hornsea Beach, East Riding of Yorkshire, England

I SPOTTED THIS PEBBLE

AT ....

ON ....

White Granite

There's an awful lot of granite in Scotland –
Aberdeen isn't known as the Granite City for nothing
– but if you've ever watched curling at the Winter
Olympics, there's a good chance the stones you've
seen being shepherded down the ice were quarried
on the tiny Scottish island where this granite pebble
was found.

This pebble is a white granite with glassy grey
quartz, white feldspar and black mica. It has
interlocking crystals that make it very hard and
resistant to water, and it comes from the remnants of
an ancient volcano.

All that's left of the volcano now is the distinctive
conical island of Ailsa Craig. It's uninhabited, and a
bird sanctuary these days, but you can still visit it by
boat and explore the beach for yourself.

Found: Ailsa Craig Island, Firth of Clyde, Scotland

I SPOTTED THIS PEBBLE

AT ....

ON ....

Glossary

Amygdale/ amygdaloidal Minerals, such as quartz, calcite and zeolite, filling a hole (*vesicle*) in an *igneous* rock.

Andalusite Aluminium *silicate* mineral with elongated rectangular crystals, brown, red or olive-green in colour, found in *metamorphic* rocks.

Augite *Silicate* mineral with green, brown or black-coloured squarish crystals.

Basement The oldest crystalline rocks underlying the continents, usually made of *igneous* and *metamorphic* rocks.

Biogenic Formed by living things.

Biotite See *mica*.

Calcite Common mineral form of calcium carbonate, often white or transparent, forms pointed crystals or a mass of fine crystals; main mineral in limestone.

Carboniferous The geological time period from 345 to 280 million years ago.

Chert See *silica*.

Chlorite Green clay mineral that forms flat flaky crystals similar to *mica*.

Clay	Mineral particles smaller than 0.002mm (2 microns).
Concretion	A hard rounded mass of rock formed from its enclosing rock.
Cordierite	Magnesium aluminium *silicate* mineral with blue or grey squarish crystals found in *metamorphic* rocks.
Crypto-crystalline	Mineral crystals that are too small to be seen with the human eye.
Dolomite	Calcium magnesium carbonate mineral, often white, brown or pinkish in colour, with tablet-shaped crystals; main mineral in *dolostone*.
Dolostone	Sedimentary rock mostly made of the mineral *dolomite*.
Feldspar	Group of *silicate* minerals of three main types: potassium feldspar, sodium feldspar and calcium feldspar; pink, white or grey, forms elongated rectangular to short prism-shaped crystals.
Fluorite	Calcium fluoride mineral (also known as Fluorspar) with cube-shaped crystals, white, green, purple or yellow in colour, often transparent.
Galena	Lead sulphide mineral with shiny metallic-grey cube-shaped crystals.
Garnet	*Silicate* mineral with dodecahedral (12-sided) crystals, red, black or green in colour, found in *metamorphic* rocks.

Glauconite	See *mica*.
Graphite	Crystalline form of carbon, metallic shiny surface, black or dark grey in colour. Very soft – will leave black marks on paper.
Groundmass	Finer-grained minerals in an *igneous* rock with larger crystals (*phenocrysts*).
Gypsum	Calcium sulphate mineral, soft, typically white tinted brown, grey or yellow in colour; crystals are transparent rectangular (selenite), fibrous (satin spar) or flat-bladed (desert rose).
Hornblende	*Silicate* mineral with dark green or brownish-green rectangular crystals with a glassy surface that is found in *igneous* and *metamorphic* rocks.
Igneous	Rock formed from molten rock (*magma*).
Intrusion	Process where *igneous* rock is placed into other rock.
Lapilli	Fragments of volcanic rock sized between 2 and 64mm (5/64–2^{1}/$_{2}$in).
Lopolith	A saucer-shaped *igneous* intrusion.
Magma	Molten rock.
Marble	Crystalline *metamorphic* rock formed from limestone or *dolostone* and consisting of interlocking crystals of calcite or *dolomite*.

Metamorphic	Rock that has been formed by the transformation of other rocks by heat and/or pressure.
Mica	Group of *silicate* minerals with flat flaky crystals, including biotite (black, brown or dark green), muscovite (silvery white) and glauconite (green).
Mohs Scale	Scale of mineral hardness invented by Friedrich Mohs:

10. Diamond	**The hardest naturally occurring material**
9. Corundum	
8. Topaz	
7. Quartz	**Will scratch glass easily**
6. Orthoclase	**Cannot be scratched with a steel knife**
5. Apatite	
4. Fluorite	**Easily scratched with a steel knife**
3. Calcite	**Just about scratched with a copper coin**
2. Gypsum	**Can be scratched by a fingernail**
1. Talc	**The softest mineral**

Muscovite	See *mica*.
Nodule	Rounded *concretion*.
Oceanic crust	The rock layers under the oceans.
Olivine	*Silicate* mineral with olive-green to yellowish-green tablet-shaped crystals and a glassy surface.

Ophiolite	*Oceanic* crust that has been lifted up on to the land surface.
Phenocryst	Relatively large crystals set in a finer *groundmass* in an *igneous* rock; this creates a *porphyritic* texture.
Porphyroblasts	Relatively large crystals set in a finer *groundmass* in a *metamorphic* rock.
Porphyry/ porphyritic	An *igneous* rock containing lots of *phenocrysts*.
Precambrian	The geological time period from the formation of the planet to 600 million years ago.
Pyrite	Iron sulphide mineral with brassy-coloured cube-shaped crystals. Also known as iron pyrite or 'fool's gold'.
Quartz	Common crystalline form of *silica*, colourless and transparent, white or tinted pink (rose quartz), purple (amethyst), black (smoky quartz) or yellow (citrine), with elongated pointed or hexagonal prismatic crystals, and a glassy surface.
Rutile	Titanium oxide mineral with reddish, brown or black needle-like crystals.
Sedimentary	Rocks formed by the weathering and erosion of other rocks to form sediments or precipitation from water.
Septarian	Divided into sections.

Silica   Silicon oxide which forms a crystalline mineral like *quartz* or *cryptocrystalline* minerals like chert and flint.

Silicate   Mineral made with *silica* combined with other elements such as aluminium, calcium, iron, magnesium or potassium.

Sillimanite   Aluminium *silicate* mineral with elongated rectangular crystals, typically colourless or white and found in *metamorphic* rocks.

Tourmaline   *Silicate* mineral with elongated crystals with characteristic lines along its length and hexagonal ends; blue, pink/red, green or colourless.

Vesicles   Holes in *igneous* rocks caused by gas bubbles trapped when the rock cooled.

Volcaniclastic Volcanic ash and rocks deposited as a sediment similar to sandstone.

Zeolite   *Silicate* minerals typically occuring as white and fibrous crystals in *amygdales*.

Zircon   Zirconium *silicate* mineral with colourless or grey prism-shape crystals.

Index

alabaster 12–13
amygdaloidal basalt 14–15
andesite 8
art 41

basalt 8, 14–15, 16–17
'biogenic' rocks 32, 54, 68
bluestone 34
brick 18–19

calcite 20, 32, 36, 44, 54
chalk 8, 20–1
chert 22–3
clasts 30
coal 26–7, 62, 94
coccoliths 20
collecting pebbles 10, 25
concrete 28–9
conglomerate 30–1
coral limestone 32–3
cracking pebbles 10
crinoids 44

Devil's Toenail 54
diabase 34
dolerite 34–5
dolostone 36–7

dung 50
dykes 34

echinoid 38–9

finding pebbles 10, 25
flint 38–9, 42, 57
fossils 20, 32, 38–9, 44, 50,
54–5, 68–9

gabbro 8, 46–7
Giant's Causeway 14, 16
glass 48–9
glauconitic sandstone 50–1
gneiss 8, 52–3
granite 8, 72–3, 80–1, 100–1
gryphaea 54–5
gypsum 12, 20

Hadrian's Wall 34
hag stones 57
hornfels 58–9

igneous rocks 8
intrusions 34
ironstone 60–1

jet 62–3

lapilli tuff 64–5
lava 14, 16
limestone 8, 20, 32–3, 44–5, 68–9
lopoliths 46

magic 57
magma 8, 16, 34, 46
marble 8
Mars 30
metamorphic rocks 8
minerals 8–9
mudstone 8
music 54

ophiolite 92

paramoudra 42
phenocrysts 76, 80, 82, 98
phyllite 70–1
Picts 41
poetry 88–9
pores 36
porphyroblasts 70
porphyry 76–7, 82–3
pot stones 42
puddingstone 30

quartz porphyry 76–7
quartzite 78–9

rhomb porphyry 82–3
rock types 8
Romans 28, 50

safety 11
sandstone 8, 50–1, 66–7, 84–5
schist 8, 86–7
sedimentary rocks 8
septarian concretion 90–1
serpentinite 92–3
silica 22, 38, 42
sills 34
siltstone 8
skimming stones 75
slate 8, 70
stigmaria 94–5
Stonehenge 34

tarmac 96–7
trees 26, 94
tuff 64–5, 98–9

vesicles 14
volcanoes 8

white granite 100–1

Dimensions of sample pebbles

Width / Height cm (inches)

Alabaster 6.5 x 3.5 ($2^9/_{16}$ x $1^3/_8$)

Amygdaloidal basalt 5 x 4 ($1^{15}/_{16}$ x $1^9/_{16}$)

Basalt 8 x 5 ($3^1/_8$ x $1^{15}/_{16}$)

Brick 7.5 x 5 ($2^{15}/_{16}$ x $1^{15}/_{16}$)

Chalk 5.5 x 5.5 ($2^3/_{16}$ x $2^3/_{16}$)

Chert 4 x 2.5 ($1^9/_{16}$ x 1)

Coal 4.5 x 3 ($1^3/_4$ x $1^3/_{16}$)

Concrete 6 x 4 ($2^3/_8$ x $1^9/_{16}$)

Conglomerate 10 x 5 ($3^{15}/_{16}$ x $1^{15}/_{16}$)

Coral Limestone 6.5 x 6.5 ($2^9/_{16}$ x $2^9/_{16}$)

Dolerite 9.5 x 5 ($3^3/_4$ x $1^{15}/_{16}$)

Dolostone 6 x 4 ($2^3/_8$ x $1^9/_{16}$)

Echinoid in flint 7 x 5.5 ($2^3/_4$ x $2^3/_{16}$)

Flint 8 x 6 ($3^1/_8$ x $2^3/_8$)

Fossiliferous limestone 7 x 5 ($2^3/_4$ x $1^{15}/_{16}$)

Gabbro 6.5 x 4.5 ($2^9/_{16}$ x $1^3/_4$)

Glass 3.5 x 2.5 ($1^3/_8$ x 1)

Glauconitic sandstone 7.5 x 5 ($2^{15}/_{16}$ x $1^{15}/_{16}$)

Gneiss 9.5 x 5.5 ($3^3/_4$ x $2^3/_{16}$)

Gryphaea 6 x 4 ($2^3/_8$ x $1^9/_{16}$)

Hornfels 6 x 4.5 ($2^3/_8$ x $1^3/_4$)

Ironstone 6 x 4.5 ($2^3/_8$ x $1^3/_4$)

Jet 3.5 x 2 ($1^3/_8$ x $1^3/_{16}$)

Lapilli Tuff 6.5 x 5 ($2^9/_{16}$ x $1^{15}/_{16}$)

Layered sandstone 7.5 x 5.5 ($2^{15}/_{16}$ x $2^3/_{16}$)

Limestone 7 x 4 ($2^3/_4$ x $1^9/_{16}$)

Phyllite 8 x 7 ($3^1/_8$ x $2^3/_4$)

Pink Granite 6 x 5 ($2^3/_8$ x $1^{15}/_{16}$)

Quartz Porphyry 5 x 3.5 ($1^{15}/_{16}$ x $1^3/_8$)

Quartzite 4.5 x 3.5 ($1^3/_4$ x $1^3/_8$)

Red granite 7.5 x 5.5 ($2^{15}/_{16}$ x $2^3/_{16}$)

Rhomb Porphyry 5.5 x 5 ($2^3/_{16}$ x $1^{15}/_{16}$)

Sandstone 7 x 5 ($2^3/_4$ x $1^{15}/_{16}$)

Schist 6.5 x 5 ($2^9/_{16}$ x $1^{15}/_{16}$)

Septarian concretion 5.5 x 4 ($2^3/_{16}$ x $1^9/_{16}$)

Serpentinite 8 x 6 ($3^1/_8$ x $2^3/_8$)

Stigmaria 9.5 x 9 ($3^3/_4$ x $3^9/_{16}$)

Tarmac 4.5 x 3 ($1^3/_4$ x $1^3/_{16}$)

Tuff 8 x 4 ($3^1/_8$ x $1^9/_{16}$)

White granite 7 x 6 ($2^3/_4$ x $2^3/_8$)

Recommended guides

Nature Guide Rocks and Minerals
Ronald Louis Bonewitz (Dorling Kindersley)
Excellent paperback book with clear photographs of
rocks and minerals, and guides on collection and
identification.

The complete illustrated guide to rocks of the world
John Farndon (Lorenz Books)
Quite possibly the best and most comprehensive
explanation of the classification of igneous,
sedimentary and metamorphic rocks for amateur
geologists.

The Pebbles on the Beach: A Spotter's Guide
Clarence Ellis (Faber)
The classic UK pebble book published in 1953 and
recently republished in 2018.

Acknowledgements

I would like to thank the following for their support and encouragement:

My colleagues and friends at the British Geological Survey, in particular Kirstin Lemon and Gareth Farr who were as excited as me at the prospect of this book and sent me pebbles from Northern Ireland and Wales respectively. Other BGS colleagues who lent me their precious pebbles that made it into the book: Ailsa Napier (White granite), Rachel Cartwright (Stigmaria) and Richard Haslam (Serpentinite). Jo Mankelow for reviewing the book and generally humouring me as the work progressed!

My friends Caroline & Paul Wright for many happy times on beaches with me and my wife Joy, and for lending me their Echinoid-in-flint pebble.

My friends on twitter who have joined in the fun online when new pebble discoveries have been made and shared their own pebble discoveries.

Peter Taylor at Pavilion Books who was my original contact for the book and has guided me expertly through the whole publishing process.